Hands-On STEM

FOCUS ON FRICTION

by Joanne Mattern

I0886230

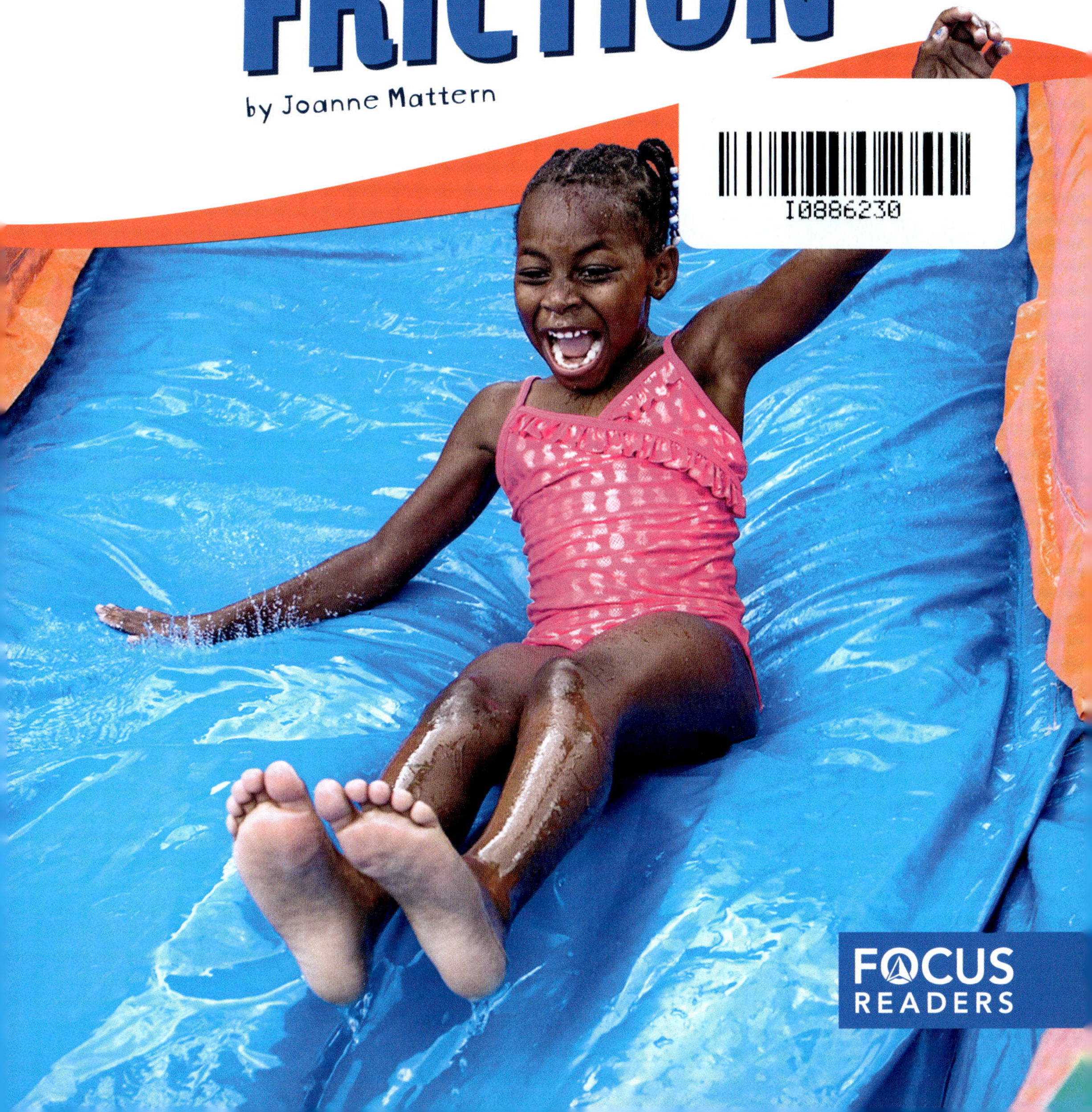

FOCUS READERS

FOCUS READERS

www.focusreaders.com

Copyright © 2018 by Focus Readers, Lake Elmo, MN 55042. All rights reserved. No part of this book may be reproduced or utilized in any form or by any means without written permission from the publisher.

Focus Readers is distributed by North Star Editions:
sales@northstareditions.com | 888-417-0195

Produced for Focus Readers by Red Line Editorial.

Content Consultant: Bruce Bolon, Associate Professor of Physics, Hamline University

Photographs ©: Brocreative/Shutterstock Images, cover, 1; goldenangel/iStockphoto, 4–5; mihtiander/iStockphoto, 7; l i g h t p o e t/Shutterstock Images, 9; bowdenimages/iStockphoto, 10–11; Casarsa/iStockphoto, 13; Leontura/iStockphoto, 15; MEROSS/iStockphoto, 17; stevecoleimages/iStockphoto, 19; puwanai/Shutterstock Images, 20–21; Avalon_Studio/iStockphoto, 23; 3DSculptor/iStockphoto, 25, 29; Red Line Editorial, 27

ISBN
978-1-63517-281-2 (hardcover)
978-1-63517-346-8 (paperback)
978-1-63517-476-2 (ebook pdf)
978-1-63517-411-3 (hosted ebook)

Library of Congress Control Number: 2017935130

Printed in the United States of America
Mankato, MN
June, 2017

About the Author

Joanne Mattern is the author of many nonfiction books for children. Her favorite subjects include science, history, and biography. She hopes to make nonfiction fun for young readers. Joanne lives in New York with her husband, four children, and several pets.

TABLE OF CONTENTS

CHAPTER 1

SLOW DOWN!

Nick straps on his helmet and steps onto his skateboard. He pushes off with his foot. At first, he rolls along quickly. But soon he starts to slow down.

Wheels help people move faster than they could by walking.

The skateboard's wheels are attached to **axles**. The skateboard slows down partly because of the friction between the wheels and the axles. Friction happens when two objects rub against each other.

Once again, Nick pushes against the street with his foot. His skateboard starts to roll faster. He is moving quickly when he sees his friend Anna. He wants to stop and talk to her. Nick puts down his foot. He drags it against the street.

The friction between the shoe and the ground helps the rider slow down.

The skateboard slows. Nick uses the friction of his shoe against the street. The friction slows his skateboard. Now he can talk to his friend.

We use friction every day. Without friction, things would slide around all the time. Friction helps people in many ways. Sometimes people look for ways to create more friction. Other times, they need ways to create less. In both cases,

Ice skates move easily across ice because they create very little friction.

friction is an important part of our lives.

CHAPTER 2

WHAT IS FRICTION?

Friction happens when one **substance** rubs against another. Solids, liquids, and gases all create friction. But some objects create more friction than others.

Friction happens when a person's body rubs against a slide.

Friction in liquids and gases is known as resistance.

Rough objects create the most friction. Suppose you are sliding a big rock across the sidewalk. It is hard to do. That's because the rock is rough. So is the sidewalk. These objects create a lot of friction when they rub against each other.

Some objects are smooth. Imagine being on ice. It is hard to start moving. But once you start, it is hard to stop. That's because the

Water resistance explains why people can't walk fast underwater.

ice is smooth. It does not create a lot of friction against your shoes.

Many liquids **reduce** friction. They can make things slippery. If you walk across a wet floor, your feet might slip. That happens because the water reduces the friction between your shoes and the floor.

A moving object has energy. When friction slows down a moving object, that energy causes the temperature to increase. For example, rub your hands together very firmly. Soon they will start

HOW FRICTION WORKS

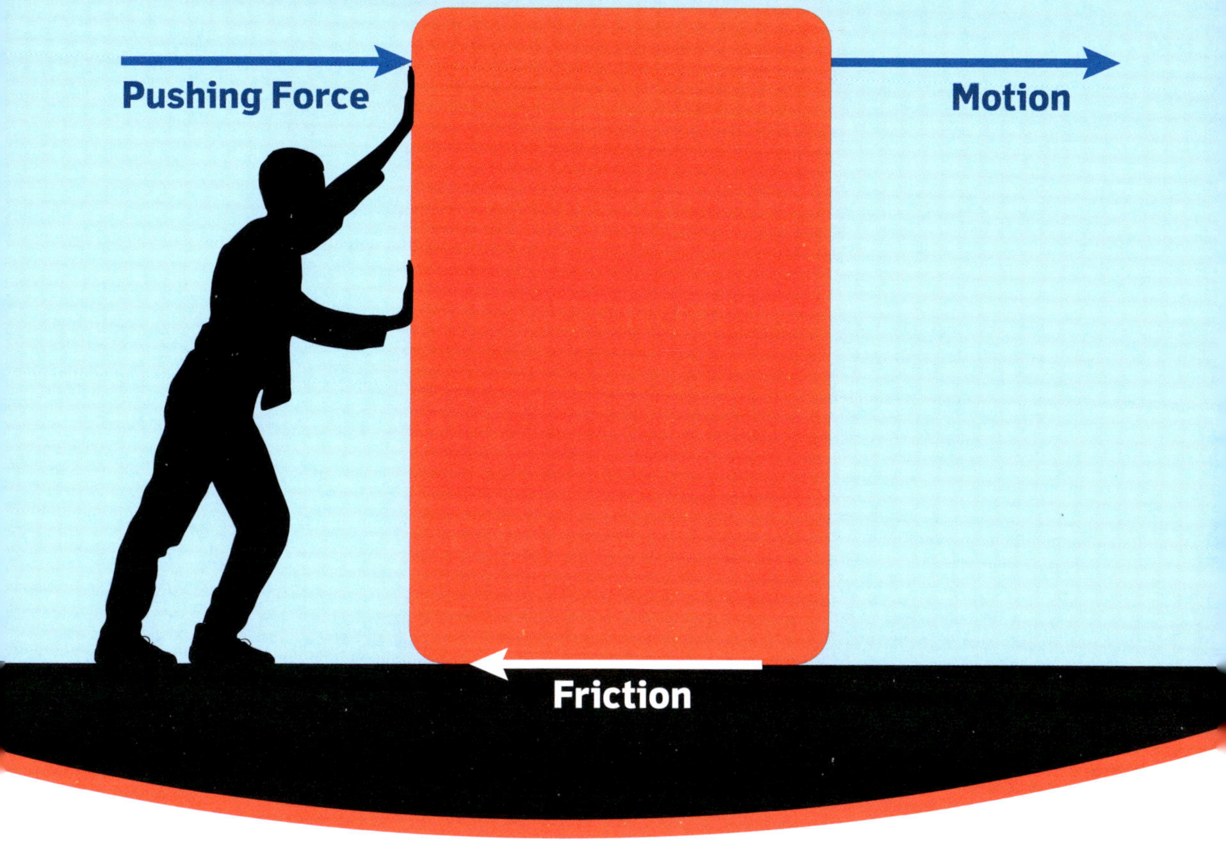

Friction happens when you push an object across a surface.

to feel warm. That is because the friction between your hands causes them to get hotter.

Friction can also cause an object to get a static charge. This happens when objects rub against one another. **Electrons** are tiny charged particles. They pass from one object to the other. Imagine walking across a carpet and touching a metal doorknob. You may get a shock. The electrons from the carpet jump between you and the metal.

Friction can be beneficial in many ways. For instance, friction helps us

A static shock happens because of friction.

slow down when we want to stop. Without friction, we would slip and slide whenever we tried to walk.

Sometimes we need to create less friction. One way to reduce the effects of friction is by using wheels. For example, it's usually easier to move a heavy box if you put it on a wagon. The friction of the rolling wheels is usually less of a problem than the friction of the box sliding along the ground.

Liquids can also help reduce friction. Adding water or oil usually makes it easier for objects to slide and move. Think about a squeaky

Friction makes large boxes difficult to push.

door. It may be difficult to close. But when you put oil on the **hinge**, the door closes more easily. That's because the oil reduces the friction at the hinge.

CHAPTER 3

FRICTION IN OUR WORLD

Friction happens in many ways. It happens whenever we move. Suppose you're riding a bike. When you use the **brakes**, they rub against the wheel. This causes the wheel to turn more slowly.

Friction helps bikes slow down.

The bike also slows down because of the friction between the wheels and the road.

Friction also happens when parts of a machine rub together. This can be bad because it causes the parts to get hot. The parts can also wear away and break. To **prevent** this, people put oil on some machines. It helps the parts move more smoothly. For example, a car engine has many moving parts. Oil helps

Car engines would get too hot without oil.

reduce the friction. It keeps the engine cooler.

Water slides also have low friction. That's because water is a liquid that has low friction.

It makes people move quickly down the slide. If the slides did not have water, people would go much slower. Water reduces the friction between the body and the slide.

The **atmosphere** around Earth is made of gases. When a rocket moves through the gases, it experiences friction. That friction causes the rocket to get very hot. For this reason, rockets are covered with special tiles. They protect the rocket from getting too hot.

When scientists design rockets, they must think about friction.

Friction is part of our world in many ways. Whenever you move or slip, think about how friction is acting on your body!

SCIENCE IN ACTION!

CREATING FRICTION

Does an object experience more friction on some surfaces than on others? Do this experiment to find out.

Step 1: Get a wooden board. It should be about 3 feet (0.9 m) long and 1 foot (0.3 m) wide. Cover one side of the board with sandpaper.

Step 2: Make sure the sandpaper side is facing up. Lift one end of the board to create a ramp.

Step 3: Place a book at the top of the ramp. Let go of the book. If the book does not slide, make the ramp steeper. See how quickly the book slides down.

The amount of friction depends on the type of surface.

Step 4: Flip the board over so that the smooth side is facing up. Keep the ramp at the same angle.

Step 5: Place the book at the top of the ramp. Let go of the book. See how quickly the book slides down this time.

Which surface creates more friction on the book? How could you make the book slide even faster?

FOCUS ON

FRICTION

Write your answers on a separate piece of paper.

1. Write a sentence that describes the main idea of Chapter 3.
2. Would you want to ride in a car that did not have enough oil? Why or why not?
3. What usually happens when friction slows down a moving object?
   - A. The object gets hotter.
   - B. Electrons are created.
   - C. Gas is created.
4. What might happen if there were no friction in the world?
   - A. Solids and liquids would turn into gases.
   - B. Moving objects would keep moving.
   - C. Objects would not be able to move.

5. What does **beneficial** mean in this book?

*Friction can be **beneficial** in many ways. For instance, friction helps us slow down when we want to stop.*

A. hot
B. slow
C. useful

6. What does **rough** mean in this book?

***Rough** objects create the most friction. Suppose you are sliding a big rock across the sidewalk.*

A. bumpy or uneven
B. easy to move
C. difficult to stand on

Answer key on page 32.

GLOSSARY

atmosphere
Gases that surround a planet.

axles
Bars that wheels spin on.

brakes
The parts of a vehicle designed to help it stop or keep it stopped.

electrons
Charged particles that can be in atoms or on their own.

hinge
An object that connects a door to a wall and allows the door to open.

prevent
To stop something from happening.

reduce
To make less.

substance
Any type of material, such as a solid, liquid, or gas.

TO LEARN MORE

BOOKS

Barchers, Suzanne I. *Friction*. Huntington Beach, CA: Teacher Created Materials, 2015.

Boothroyd, Jennifer. *Why Do Moving Objects Slow Down?: A Look at Friction*. Minneapolis: Lerner Publications, 2011.

Mullins, Matt. *Friction*. New York: Children's Press, 2012.

NOTE TO EDUCATORS

Visit **www.focusreaders.com** to find lesson plans, activities, links, and other resources related to this title.

INDEX

Answer Key: 1. Answers will vary; **2.** Answers will vary; **3.** A; **4.** B; **5.** C; **6.** A